Biogeography Lab Manual

Second edition

Mark Hecht

Front cover: A 'blue-tailed' Western Skink (*Eumeces skiltonianus*)

Biogeography Lab Manual
Second edition

Mark Hecht

First Printing: 2014

ISBN 978-1-312-79239-5

Mark Hecht Publishing
Calgary, Alberta, Canada

mhecht@mtroyal.ca

ISBN 978-1-312-79239-5

9 781312 792395   90000

Table of Contents

Taxonomy
"defining the units of life"

In order to make sense of the biota around us we need to first assign names. To assign names we must have a definable entity. The common global taxonomic system today uses the physical attributes of physical objects and their component parts to define them. Once defined, they can be given a 'scientific name' based on their physical structure, and similarity to other biota with similar physical structures. While by no means is it the only way to categorize the world around us, it is certainly the most widely accepted and understood.

This exercise will familiarize you with the taxonomy of plants.

Exercise

1. Using plant identification handbooks or other resources available to you, identify the following plants according to the binomial nomenclature system—*Genus* and *species*.

Example: **"Ponderosa Pine"**

Kingdom:	Plantae
Division/Phylum:	Coniferophyta
Class:	Pinophyta
Order:	Pinopsida
Family:	Pinaceae
Genus:	*Pinus*
Species:	*ponderosa*

"Lodge pole Pine"

Kingdom: Plantae

Phylum: Coniferophyta

Class: _____

Order: _____

Family: _____

Genus: _____

Species: _____

"Western Red Cedar"

Kingdom: _____

Phylum: _____

Class: _____

Order: _____

Family: _____

Genus: _____

Species: _____

"Choke Cherry"

Phylum: Anthophyta

Class: Dicotyledons

Order: _____

Family: _____

Genus: _____

Species: _virginiana_

"Saskatoon"

Phylum: _____

Class: _____

Order: Rosales

Family: Rosaceae

Genus: _Amelanchier_

Species: _alnifolia_

"Yarrow"

Class: Dicotyledon

Order: Asterales

Family: _____

Genus: _Achillea_

Species: _____

"Oregon Grape (Barberry)"

Class: _____

Order: _____

Family: _____

Genus: _____

Species: _aquifolium_

2. Give the Latin name (***Genus*** and ***species*** *only*) for the plants below that have been labeled with their "common name". Remember that common names can be confusing because:

...the same common name can apply to different plants in different geographic regions.

Example: Ironwood can refer to a bush in Paraguay or to *Spirea* in British Columbia

...numerous common names can exist for one plant.

Example: Lithospermum ruderale can be called Lemonweed, Stoneseed, Puccoon, Wooly Gromwell, or Western Gromwell. Not only can various common names be found in the different geographic regions within the same regions, different groups of people may prefer one common name over another. *Lithospermum ruderale* in southern BC is often called Gromwell by older generations, but younger generations often say Lemonweed

According to the common names given below use any reference books or other sources available to find the scientific name for each. Also include habitat description.

Trembling Aspen   __*Populus tremuloides*___   __*forests_and_parklands*_____

Engelmann Spruce   _____   *subalpine,_Rocky_mountains*__

Shrubby Cinquefoil   _____   _____

Wolf Willow   _____   _____

Soopalallie (can also be called: _____)

_____   _____

Oregon Grape   _____   _____

Red-osier Dogwood   _____   _____

Kinnikinnick (can also be called: _____)

_____   _____

Common Juniper   _____   _____

Prairie Crocus   _____   _____

Mountain Arnica   _____   _____

Canada Goldenrod _____ _____

Black-eyed Susan _____ _____

Brown-eyed Susan (can also be called: _____)

_____ _____

Canada Thistle _____ _____

Burdock _____ _____

Pasture Sage _____ _____

Stinging Nettle _____ _____

Yellow Hedysarum _____ _____

Blue-eyed grass _____ _____

Northern Valerian _____ _____

Purple Clematis _____ _____

Tufted White Prairie Aster_____ _____

Prickly Pear Cactus _____ _____

Fireweed _____ _____

Indian Pipe _____ _____

Wood Lily _____ _____

Glacier Lily _____ _____

Indian Hellebore _____ _____

Blue Gramma _____ _____

Crested Wheatgrass _____ _____

Cheatgrass (can also be called: _____)

_____ _____

Kentucky Bluegrass _____ _____

Cattail _____ _____

Rose (Alberta Wild) _____   _____

Rough Fescue   _____   _____

Sweet Grass   _____   _____

Sticky Purple Geranium _____   _____

Lowbush Cranberry   _____   _____

Flax   _____   _____

3. With the following plants, identify their scientific names and also note any interesting points about medicinal uses, edible qualities, mythological connections or other points that various authors have noted in their reference books.

Common Name	*Scientific Name*	*Edible?*		*Other Notes*
Red Raspberry	_____	Y	N	_____
Lemonweed	_____	Y	N	_____
Asparagus	_____	Y	N	_____
Blueberry	_____	Y	N	_____
Prostrate Amaranth	_____	Y	N	_____
Baneberry	_____	Y	N	_____

a. Which one of the previous plants is **NOT** 'native' to North America?

_____

b. Name at least one of the plants whose species **IS** native to North America **AND** also native to at least one other continent?

Species?   _____

Continent?   _____

c. What similar conditions might allow a species to exist in different regions of the world? _____

_____

Notes:

Distribution of Life
"Where and why there?"

In the distribution of life we find that key physical factors not only limit the range of species but can also influence a temporary change in species characteristics and even influence the process of speciation itself.

The three main physical factors are: Temperature, Moisture, and Light.

This exercise will familiarize you with the physical environment and the role that various factors have on the distribution of biota.

Bergmann's Rule: (1847)
Heat loss of an organism is directly proportional to its surface-to- volume ratio.

Allen's Rule:
Shorter the extremities are relative to body mass, the lower the rate of heat loss.

Exercise

Betancourt & Smith (1998) analyzed middens and coprolites from Bushy-Tailed Woodrats on the Colorado Plateau. Their research specimens came from five locations and covered a time span from 20,000 years before present (BP) to approximately 2000 years BP. This dates from the late Pleistocene Epoch through most of the Holocene Epoch. Global temperatures are known to have fluctuated significantly over this time period.

1. Plot the selected data of Body Mass versus Temperature on the following page.

BUSHY-TAILED WOODRATS (*Neotoma cinerea*)
BODY MASS (g) in Relation to MEAN JULY TEMPERATURE

July. Temp.	18.4	21.1	21.5	21.9	15.9	22.0	22.2	21.9	20.6	17.0	19.4	20.1	20.2
Body Mass	425	360	345	325	465	260	280	325	375	450	405	390	385

Selected data from: Betancourt, J.L. & Smith, F.A. (1998) Response of Busy-Tailed Woodrats (*Neotoma cinerea*) to late Quaternary Climatic Change in the Colorado Plateau. *Quaternary Research* **50**, 1-11.

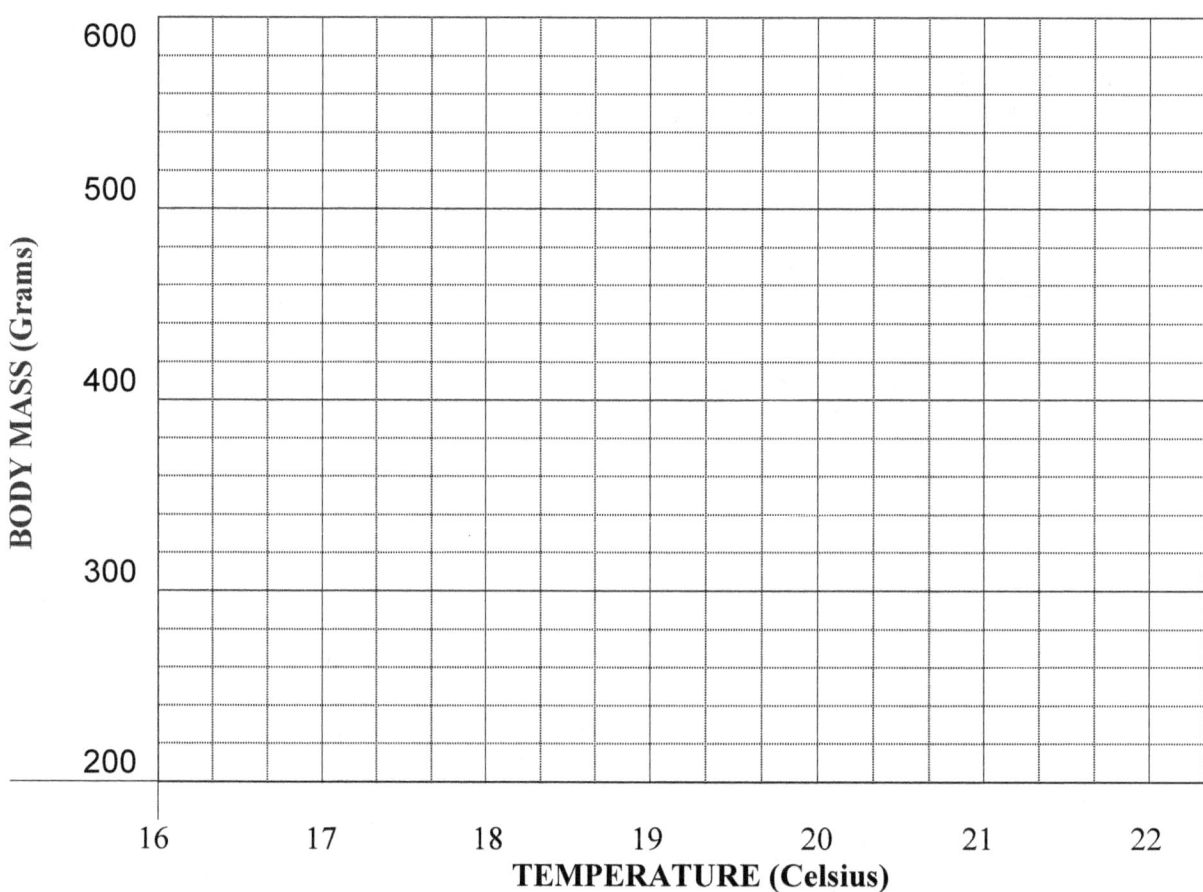

2. What can you conclude about the correlation between the Body Mass of Bushy-Tailed Woodrats and changing temperatures?

_____

3. If Body Mass and Temperature are correlated, what are two hypotheses you could suggest if mean annual temperature dropped significantly?

_____

_____

4. While Temperature, Moisture and Light are key factors in determining the distribution of species, what other factors might be important?

_____

5. If all physical factors are favourable to a specific species and:

a. ...that specific species is <u>NOT</u> found in a particular place, WHY might that be?

_____

_____

_____

b. ....that specific species <u>IS</u> found in a particular place, what other factors may limit its increase in numbers?

_____

_____

_____

Notes:

Elements & Minerals
"another limiting factor"

Physical and Biological Limiting Factors determine where an organism can be extant; where it can live. A key factor is soil chemistry and more specifically, the questions around what element, found in combination with other elements to form minerals, the soil can supply to an organism.

A single element can be deficient or toxic, acutely or chronically, and can lead to health impairment that may ultimately determine an organism's viability in that location.

In this exercise you will choose an element and become familiar with the role it plays in human health.

Exercise

Choose one of the elements below for further study in the following questions

Cadmium	Manganese
Chromium	Mercury
Cobalt	Molybdenum
Copper	Nickel
Fluorine	Silver
Iron	Vanadium
Lead	Zinc

1. What is considered to be a safe level in humans according to a reputable source? Reputable sources include various bodies but a good start is either the US Environmental Protection Agency or Health Canada. Note that what is considered an acceptable level has a lot to do with the metrics being used. Different criteria for measuring acceptable levels can include blood toxicity, muscle tissue content, and saliva content and even hair analysis. Figure out which criteria is a good metric for the element you have chosen. Consider that many of these elements are rarely isolated and found not as elements, but instead, as minerals. You will need to determine the impact of the element you have chosen and consider its most common positive or negative effect on human beings.

   Acceptable maximum level: _____

   According to: _____

2. What are the human symptoms of toxicity, if any?

3. What are common sources of toxicity?

4. What are the human symptoms of deficiency, if any?

5. What are the common reasons for deficiency?

6. What are food sources for this element?

7. Is this element found equally around the world or do some regions suffer from excess/deficiency? Where?

Disturbance & Succession
"recognizing the inevitability of change"

Disturbance is a regular part of any landscape spatially and temporally. Various theories and models have suggested discernible patterns concerning regeneration of the biological landscape. However, exceptions and unexplainable circumstances leave room for continued work on refining these models.

Exercise

1. Find a scientific article that deals with "succession."

2. Cite the article you found:

3. What is the hypothesis in the article?

4. Does the author mention any of the models below?

      Clement's Succession Model   ____

      Tolerance Model   ____

      Inhibition Model   ____

      Random or IFC Model   ____

5. What are your own thoughts about the article you found?

Dispersal, Colonization, & Invasion
"movement brings change"

While geographic isolation can lead to immense divergence and speciation, resulting in an increase in biodiversity, that isolation is never absolute. When species come to newly occupy geographic spaces, we often find patterns emerging.

This exercise introduces you to quantitative techniques that attempt to establish patterns of that movement and colonization.

Colonization:
When a propagule arrives in an area previously unoccupied by the species and establishes a reproducing population, leading to range expansion.

Exponential Population Growth
A population trajectory uninhibited by limited food resources or biological factors. Represented on a graph as an infinite upward curve

Logistic Population Growth
A population growth of a colonizing species in a setting with finite resources. Represented on a graph as an "S"-curve.

Carrying Capacity Overshoot
When a population experiences growth beyond the average carrying capacity of the area, and experiences a catastrophic decline back to, or below, the original carrying capacity.

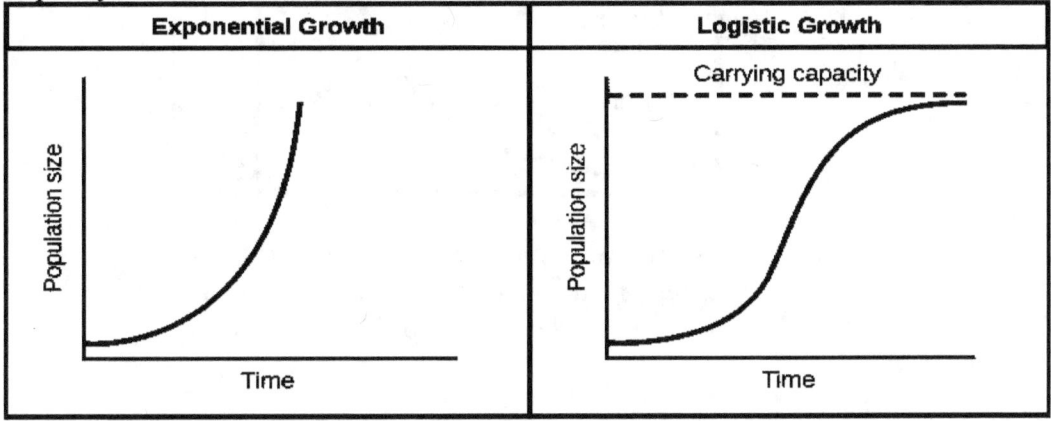

Exercise

In 1944, the US Coast Guard placed 29 Woodland Caribou *(Rangifer tarandus)* on the isolated Bering Sea island of St. Matthew, Alaska. There were 24 females and 5 males. By 1957, the population exploded to 1350. Six years later, in the summer of 1963, the population had continued to increase to an estimated 6000 Caribou.

However, in the winter of 1963/64, the Woodland Caribou went through a massive die-off and by the next summer in 1964, less than 50 remained. In 1966, another count was again conducted. The result was a total of 42 Caribou; all female with the exception of one older male. By the 1980s, the introduced Woodland Caribou population on St. Matthew Island had vanished.

Data from: Klein, D. R. (1959) Saint Matthew Island reindeer-range study, U.S. Fish and Wildlife Service Special Science Report: Wildlife, 43, p.48.

1. Review the graph below. Does the graph appear to represent "Exponential Growth" or "Logistic Growth"? _____

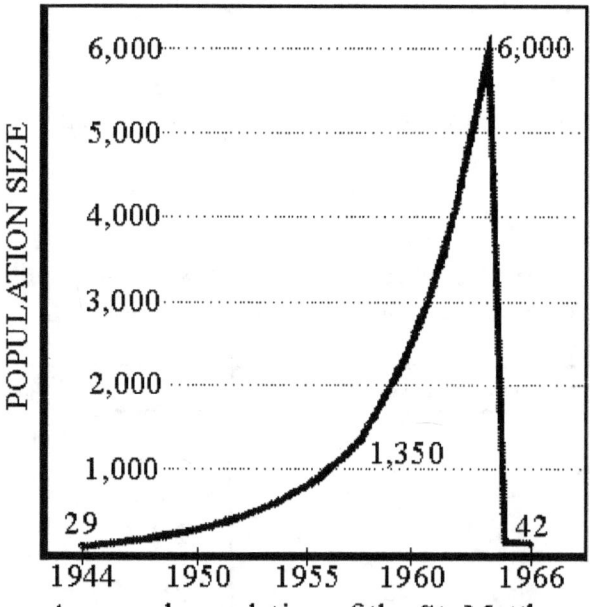

Assumed population of the St. Matthew Island reindeer Herd. Actual counts are indicated on the population curve.

2. Biogeographers often use mathematical modeling techniques to interpret or predict population changes. Using the formula below, figure out the TIME TO DOUBLE (Td) for the years from 1944 (29 Caribou) to 1957 (1350 Caribou). You will also need to use Logarithms to arrive at your answer.

$$\text{New value} = \text{initial value} \times 2^{t/Td}$$

Td (Time to double) = _____ **years**

3. Knowing the **TIME TO DOUBLE (Td)**, estimate how many Caribou there should mathematically be in the year 1963? (Note: Calculate from 1957 to 1963)

$$\text{New value} = \text{initial value} \times 2^{\,t/Td}$$

Caribou in 1963 = _____ number (estimated)

4. What might account for the <u>actual</u> number or 6000 Caribou on ST. Matthew Island in 1963, being <u>less</u> than your estimate from the previous question?

5. Does the ACTUAL number of Caribou from 1944 to 1963 more closely resemble "Exponential Growth" or "Logistic Growth"?

_____

6. The quick rise and decline in population of *Rangifer tarandus* on St. Matthew Island is an example of:

_____

Notes:

Biodiversity
"how many?"

Biodiversity is often seen as an indicator of ecological stability. However, the spatial and temporal patterns of biodiversity vary significantly. Applying a quantitative approach may offer clues as to whether or not an ecosystem is "normal' in comparison to equivalent areas, although no general rule or theory exist for predicting species distribution within that ecosystem

This exercise gives you a 'snapshot' of biodiversity evenness among avifauna in the City of Calgary. Various quantitative techniques exist for measuring and establishing biodiversity values. While these are noted below, we will only examine species evenness from a quick and general perspective.

Species Richness
The number of different species in a given area. The terms 'species richness' and 'biodiversity' are 'often used interchangeably in large-scale studies. An increase in area studied usually leads to an increase in species richness.

Species Evenness
The degree to which the number of different organisms are evenly divided between the different species.

Species-Area Curve
Generally, species richness increases with area BUT NOT at a linear rate. This nonlinear Species-Area Curve is calculated as:

$$S = cA^2$$

Pielou's Evenness Index
Named after the Canadian statistical ecologist, Evelyn Chrystalla Pielou, this index gives a J' value between 0 and 1, with values closer to 1 indicating more evenness. H' and H'_{max} come from the values calculated in the Shannon Diversity Index.

$$J' = H'/H'_{max}$$

Shannon Diversity Index

$$H' = -\sum_{i=1}^{R} p_i \log p_i$$

Simpson's Index of Diversity

Another measure to calculate Species Evenness and the calculation we will use for the exercise below. The index is as follows:

Index =   1-D

Where D = $\frac{\sum n(n-1)}{N(N-1)}$

Example

Species	Number (n)	n(n-1)
Woodrush	2	2
Holly (seedlings)	8	56
Bramble	1	0
Yorkshire Fog	1	0
Sedge	3	6
Total (N)	15	64

Putting the figures into the formula for Simpson's Index.....

D = $\frac{\sum n(n-1)}{N(N-1)}$

D = $\frac{64}{15(14)}$

D = $\frac{64}{210}$

D =   0.3

Then:

Simpson's Index of Diversity: 1 - D = **0.7**

Exercise

Every year volunteers in cities around the world do Christmas bird counts. Using the data provided from the City of Calgary, 2011 CBC Christmas bird count, you will determine Species Evenness for the city's winter bird population.

Data source: www.naturecalgary.com/wp-content/uploads/2012/06/Calgary-CBC-Team-Report-Sheet-2011-Final.pdf

Species	Individuals	Species	Individuals
Cackling Goose	2	Northern Flicker	142
Canada Goose	8186	Pileated Woodpecker	1
Wood Duck	7	Northern Shrike	2
American Widgeon	2	Blue Jay	102
Mallard	14623	Black-billed Magpie	2384
Northern Pintail	1	American Crow	36
Lesser Scaup	1	Common Raven	213
Canvasback	1	Black-capped Chickadee	1358
Pine Siskin	571	Boreal Chickadee	9
House Sparrow	6214	Red-breasted Nuthatch	347
Bufflehead	178	White-breasted Nuthatch	55
Common Goldeneye	3062	Brown Creeper	12
Barrow's Goldeneye	20	Hoary Redpoll	13
Hooded Merganser	5	Golden-crowned Kinglet	18
Common Merganser	60	Townsend's Solitaire	8
Gary Partridge	125	American Robin	61
Ring-necked Pheasant	24	European Starling	515
Ruffed Grouse	3	American Pipit	1
Pied-billed Grebe	1	Bohemian Waxwing	19593
Bald Eagle	19	Cedar Waxwing	10
Northern Harrier	1	American Tree Sparrow	6
Sharp-shinned Hawk	8	Harris's Sparrow	2
Cooper's Hawk	6	Dark-eyed Junco	213
Northern Goshawk	6	Snow Bunting	3
Accipiter sp.	2	Rusty Blackbird	1
Red-tailed Hawk	1	Red Crossbill	139
Rough-legged Hawk	14	Gray-crowned Rosy-Finch	1
Merlin	18	Pine Grosbeak	862
Gyrfalcon	2	Purple Finch	2
Prairie Falcon	2	House Finch	1280
Killdeer	5	White-winged Crossbill	1129
Gull sp.	1	Common Redpoll	1526
Rock Pigeon	3221		
Great Horned Owl	5		
Belted Kingfisher	1		
Downy Woodpecker	131		
Hairy Woodpecker	24	**TOTAL Individuals/ 69 species**	66599

1. Determine the top 10 most abundant species, <u>not all 69</u>, and fill in the table below.

Species	Number (n)	n(n-1)
1.		
2.		
3.		
4.		
5.		
6.		
7.		
8.		
9.		
10.		
Total (N)		

2. Total number of birds for <u>all 69</u> species: _____

3. Species Evenness among <u>top 10</u> bird species. _____

Calculate D using: $\dfrac{\sum n(n-1)}{N(N-1)}$

a. If **D =** _____, then **1-D** gives a Simpson Index Diversity of: _____

b. If we had included all 69 species of birds in our calculations, how would species evenness likely change?　　　　　　　　Increase / Decrease

c. Species Evenness among <u>all 69</u> bird species.

How many birds are in the top 10% (first 7 species from your chart): _____

Total number of birds among all 69 species: _____

Percent: _____%

If the top 7 species (10% of total) constitute _____% of total birds counted, could we consider this to be even? _____

4. From the previous example we find that only a few bird species make up a large proportion of the total bird population in the City of Calgary during the winter season. Biogeographers and ecologists often find this same pattern among populations worldwide.

a. On the following graph you will plot the bird species distribution for the City of Calgary during the winter of 2011. Divide your species count into seven equal groupings. Since there are 69 species, make your first group the total number of individuals from the first nine most abundant species. Make each of the remaining six groups, the sum of ten species in sequential order of abundance. Plot as a histogram. The first group has been started for you.

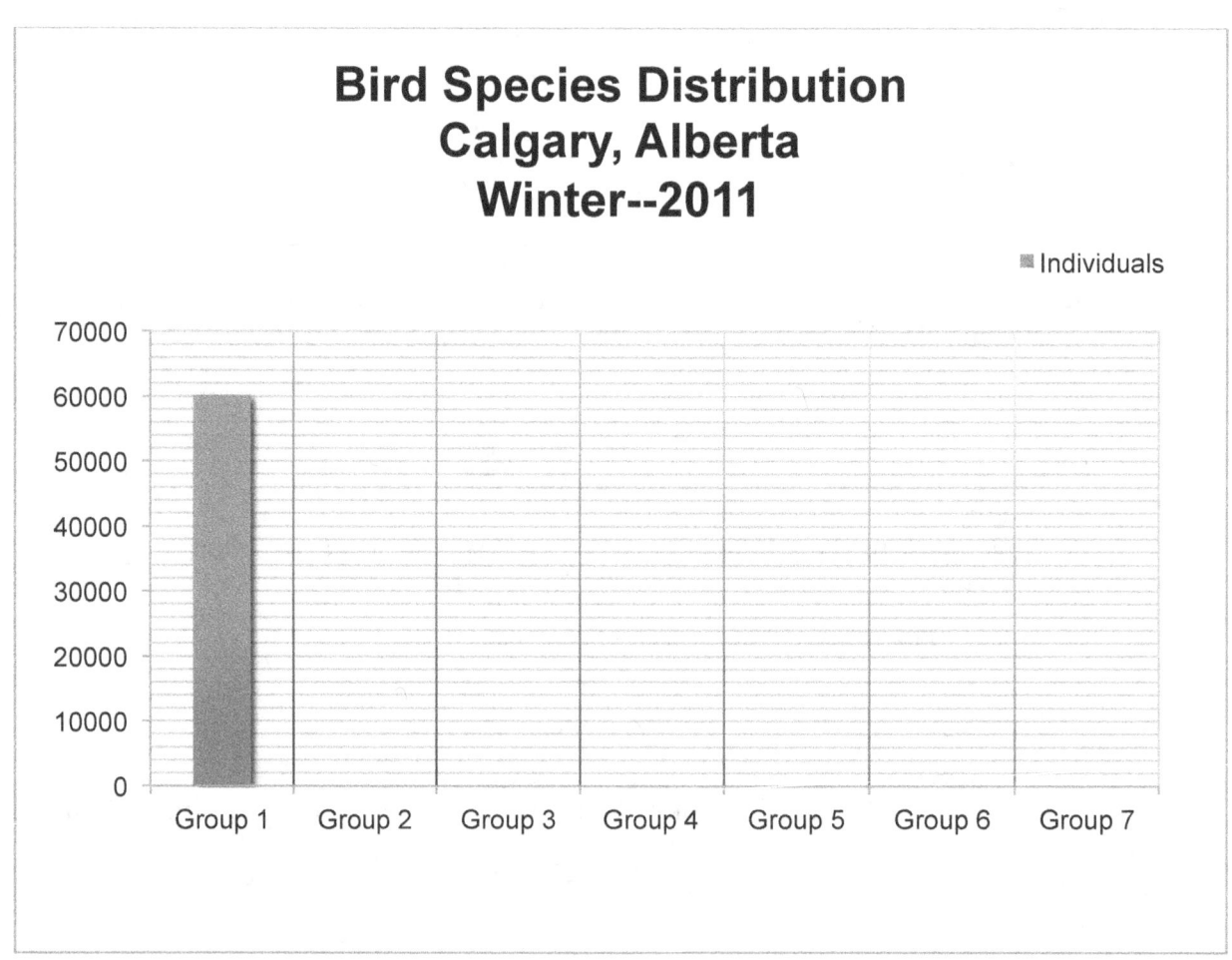

b. What is this type of curve (distribution) called?

_____

Island Biogeography
"a magical prescription"

The *Equilibrium Theory of Island Biogeography* initially seemed to offer a quantitative means for determining a stable value for species richness both spatially and temporally. For conservationists, it finally offered a clear formula that could be used for allocating resources effectively. Limitations to the theory soon became apparent though. In the end it was revealed that once again we have no reliable or definitive means of forecasting species richness. The *Equilibrium Theory of Island Biogeography* however, has become the foundation for further theoretical thought and a guide in continuing research efforts.

This exercise will familiarize you with the *Equilibrium Theory of Island Biogeography*.

"Getting there"- Immigration

Lomolino et al. (1989) studied the number of mammal species found on isolated mountain peaks in the southwest United States.

1. According to their graph below, what can we say about species richness in relation to mountain peak isolation? Think about mountain peaks as *'islands'*.

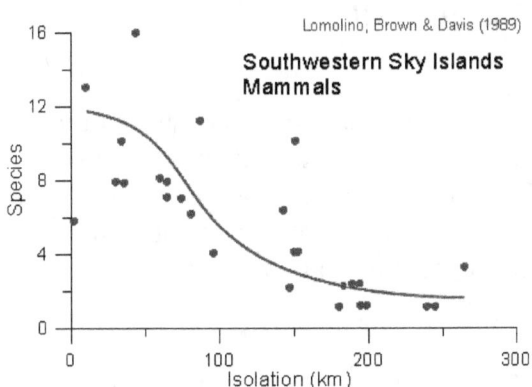

Source: M. Lomolino, J.H. Brown, and R. Davis (1989) Island Biogeography of Montane Mammals in the American Southwest, *Ecology*, Vol.70, No.1, p.187.

"Does size matter there"?

	Area (km^2)	Angiosperm Genera	Bird Genera
Solomon Islands	40 000	654	126
New Caledonia	22 000	655	64
Fiji Islands	18 500	476	54
New Hebrides	15 000	396	59
Samoa Group	3 100	302	33
Society Islands	1 700	201	17
Tonga group	1 000	263	18

Cox, C.B. & Moore, P.D. (2005) Biogeography, An Ecological and Evolutionary Approach, 7th Ed.Oxford, UK: Blackwell Publishing: 171

2. Examine the chart above. What can we say about island area and number of species?

_____

"Dying there"- Extinction

Local rates of extinction are hard to measure, since immigration may counter the local extinction (a.k.a. the "Rescue" effect) and stochastic extinction may be more common on smaller islands than large. However, MacArthur & Wilson (1967) suggested with their _Equilibrium Theory of Island Biogeography_ that extinction rates would generally increase as immigration rates decrease, thus reaching an 'equilibrium point' where the number of species present would stabilize.

3. Examine the chart overleaf displaying the species that were present on the island of Rakata recorded from the 1883 explosion of Krakatau onwards. Circle the best answer in the following questions.

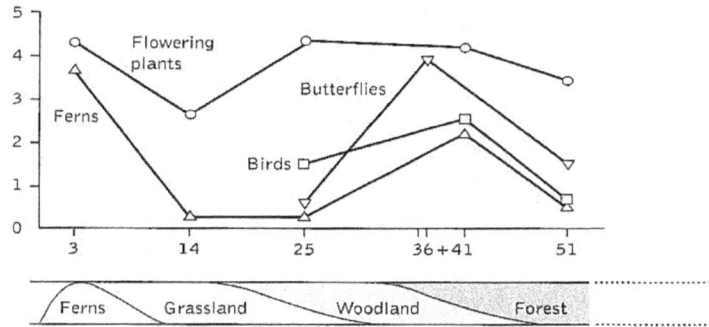

Source: C.B. Cox & P.D. Moore (2005) Biogeography, An Ecological and
Evolutionary Approach, 7th Ed, Oxford: Blackwell Publishing, p.181.

a. For *Flowering Plants*, is there an overall <u>increase</u> or <u>decrease</u> in the immigration rate (YR3 vs. YR51)?

b. For *Birds*, is there an overall <u>increase</u> or <u>decrease</u> in the immigration rate?

c. For *Ferns*, is there an overall <u>increase</u> or <u>decrease</u> in the immigration rate?

d. For *Butterflies*, is there an overall <u>increase</u> or <u>decrease</u> in the immigration rate?

    i. What specific environment is a rate increase associated with?

_____

    ii. What specific environment is a rate decrease associated with?

_____

e. While the average immigration and extinction rate is portrayed in the graph below, consider how the immigration rate would change based on distance from a continent. Draw in two new immigration lines. One will represent **a)** a large island NEAR the continent, and the second will represent **b)** a small island far from a continent.

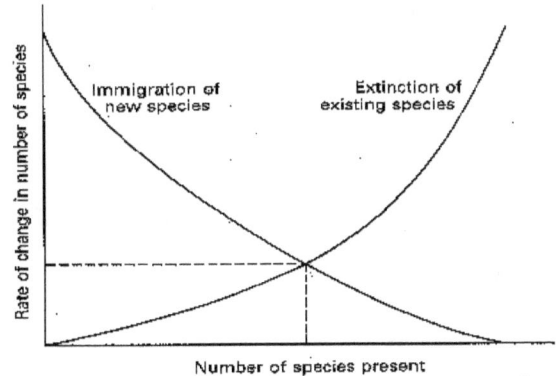

"Second thoughts on the Theory"

4. According to the data from the graph in Q.3, does immigration always occur at a constant declining rate for all species?

_____

5. What is the generally accepted scientific consensus today about the relationship between island size and "species equilibrium"?

_____

Biogeographic Boundaries

"the big picture"

"When considering vegetation on a global scale, a classification by physiognomy (general form and lifestyle) of the vegetation is more helpful than the use of taxonomy in the comparison of similar climatic locations in different parts of the world."

Cox, C. & Moore, P. (2005) Biogeography, An Ecological and Evolutionary Approach. Malden, Massachusetts: Blackwell.

Flora and fauna have long shown global patterns in geographic groupings. However, debate about the criteria on which to base those groupings has never been fully resolved, as evidenced by the quote above.

This exercise will familiarize you with the metrics and associations that are used to define communities of flora and fauna at a global scale. It will also introduce you to the evidence Alfred Russel Wallace used, that lead him to propose in 1859 a faunal boundary between what he termed the Indo-Malayan Region and Australo-Malayan Region. The boundary became known as "The Wallace Line" and it is still used today as a general division between major global bioregions.

Species-Area Relationship
Larger areas tend to support more species

Oceanic Island
An island, geologically formed in isolation

Continental Island
An island with original connections to a continental land mass

Local Sea Level
The height of the sea in relation to a land benchmark, averaged over many years

Eustatic Sea Level
The changing height of global sea levels, as ocean waters either increase or decrease in volume, usually as a result of water storage or release in glaciers during climatic alterations over hundreds or even thousands of years.

Adaptive Radiation
Evolutionary diversification of a species from a single lineage into multiple species.

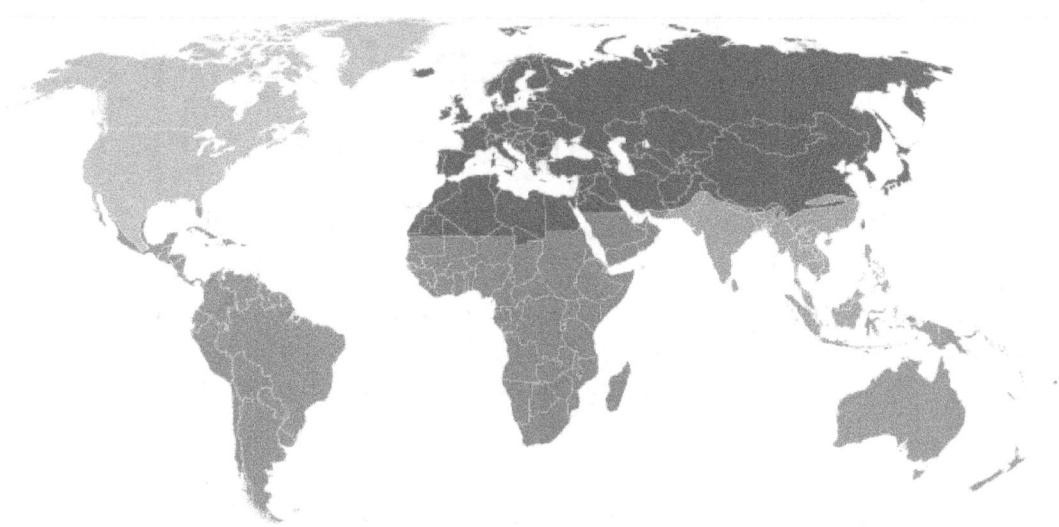

Bioregions of the World
Source: World Wildlife Fund for Nature

Bioregions (Ecozones)

As illustrated in the map above, the world can be divided into categories called Bioregions, also known as Ecozones. These divisions delineate the broadest biogeographical divisions based on the distributional patterns of all terrestrial organisms. They are based on the evolutionary history of the organisms found within the Bioregion's boundaries.

1. The terms below are the names of the five Bioregions. Find which area/continent of the world each term is roughly associated with.

Nearctic _____

Neotropic _____

Palearctic _____

Afrotropic _____

Indomalayan _____

Australasia _____

ice sheet and polar desert
tundra
taiga
temperate broadleaf forest
temperate steppe
subtropical rainforest
Mediterranean vegetation
monsoon forest
arid desert
xeric shrubland
dry steppe
semiarid desert
grass savanna
tree savanna
subtropical dry forest
tropical rainforest
alpine tundra
montane forests

Biomes of the World

Biomes

Often referred to as *Ecosystems*, these spatial categories represent climatically and geographically contiguous areas. They are often considered as similar habitat types. Similar physiological and morphological responses of the flora and fauna are used to define the biome. Examples of physiological and morphological response include plant structure, leaf-shape and spacing. Climate is also used to define biomes. Genetic and evolutionary relationships are not used.

What is the difference between Biomes and Bioregions?

Biomes are associated with habitat and the flora and fauna found within those habitat types—habitat and climate defines the geographic boundaries.

Bioregions of the world start with the premise that flora and fauna will have similar evolutionary histories contained within a broadly defined geographic area—geography defines the evolutionary boundaries.

For the following questions, we will deal only with Bioregions.

How are boundaries determined between Bioregions?

Trying to determine the spatial boundaries between Bioregions poses some major difficulties. For example, the following initial questions must be considered:

- *What will the boundaries apply to?* Flora, or fauna, or perhaps both?
- *What criteria will be used?* Existence of herbivores? Absence of a certain genus or family? Endemism? Genetic and phenotypic characteristics?

With many answers to these, and many other potential questions, various proposals have been made, beginning from the mid-nineteenth century and continuing into the present. In spite of the many variations, we see enough commonalities in the location of global bioregional boundaries that a general consensus has formed around the location of major divisions in the world's flora and fauna. The map of the Bioregions at the beginning of this question set, is a contemporary example of the current boundaries of flora and fauna combined, as determined by the World Wildlife Fund.

An example of how these boundaries have historically been determined over time dates back to people such as Philip Lutley Sclater, Salomon Muller and Alfred Russel Wallace. Alfred Wallace, in 1868, proposed what would later become known as the 'Wallace Line' of which our current boundaries closely follow.

The next questions will lead you through the evidence that Wallace used in order to propose his 'Wallace Line' as a boundary between what he termed the Indo-Malaysian region and the Austro-Malaysian region.

Case study: 'The Wallace Line'

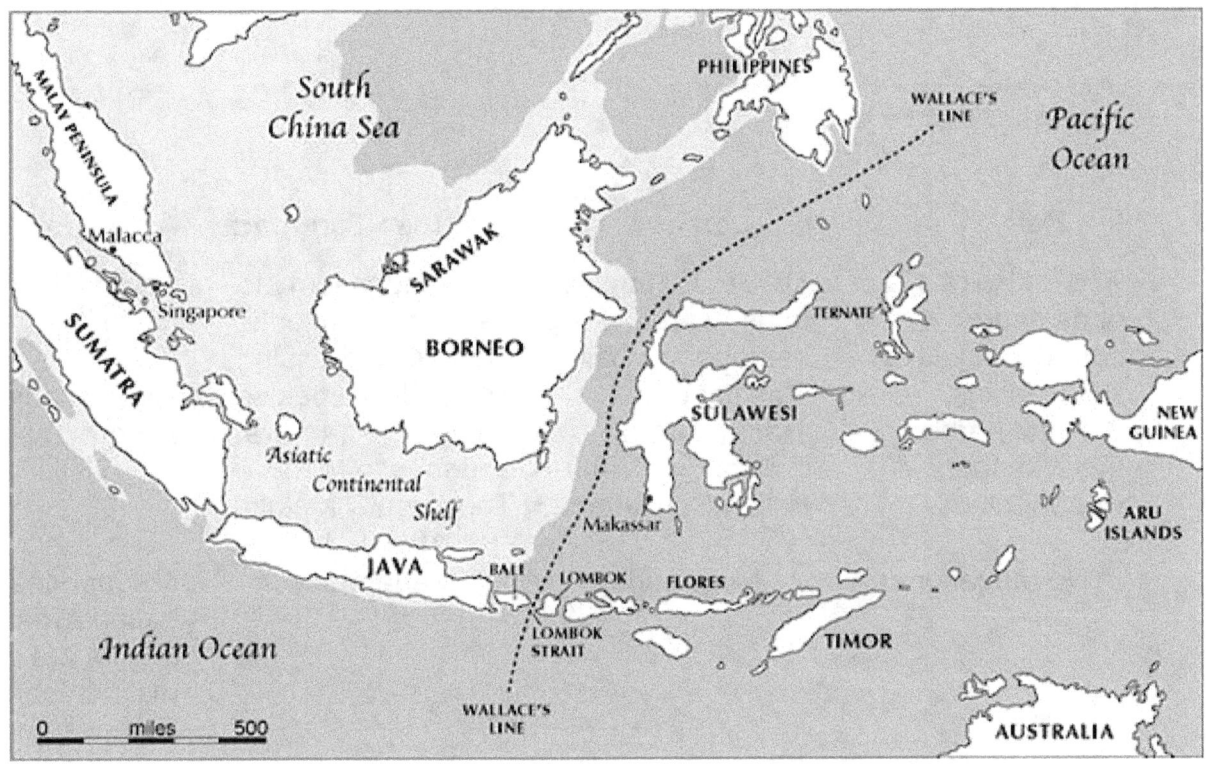

In 1856 Alfred Russel Wallace took a circuitous trip from Singapore, through Bali, then Lombok, on his way to Sulawesi. He had never intended to visit Bali. Fortuitously, he was forced by lack of transportation options to make a stopover on the island. The distance between Bali and Lombok is a mere 24 kilometers. Being a keen ornithologist and commercial collector of avifauna specimens, he immediately recognized the cacophony of bird sounds on Bali and Lombok was starkly different. On Bali, the bird sounds consist of the rata-tat-tat of woodpeckers and tonk, tonk, tonk of barbets. On Lombok the cooing of friarbirds and melodious honeyeaters heightened Wallace's curiosity.

Review the map above, which shows Wallace's proposed division between Insular Southeast Asia (Indo-Malayan Region) and Australasia (Australo-Malayan Region). The islands of note are Java, Bali, Lombok, Nusa Tenggara (south of Flores) and Timor.

Look at the map and also examine the chart overleaf to answer the following questions.

2. Of the following three islands: Java, Bali, and Lombok, which two must have been connected at some point in the past, allowing for ease of movement between the two? This would include easier movement of avifauna that are short-distance fliers and that would even include flightless birds.

_____ and _____

Bali	Lombok	Nusa Tenggara	Timor
97% of Bali's 172 birds shared w neighboring Java	50% of Lombok's birds shared w neighboring Bali	High endemism / located in the middle of the island chain	Largest island in chain
Key species: Oriental barbets, fruit thrashers, & woodpeckers.	Key species: Honeyeaters, Friarbirds, & Australian Cockatoos.	Of 562 bird species in region – 144 are endemic to Nusa Tenggara & Maluku (16%)	Highest number of endemic species – (23%)
<u>No</u> Megapodes	Megapodes	Megapodes	Megapodes + Australian avifauna origins are more distinct than those of Javanese origin.

a. Nusa Tenggara is found in the middle of the island chain. What would likely account for its high percentage of endemic species?

_____

b. Timor is the largest island in the regional chain of the Malay Archipelago with the highest percentage of endemic bird species.

i. What does this suggest in relation to Nusa Tenggara?

_____

ii. Look at the chart. Was Timor attached to a continent in the past, and if so, which continent?

_____

3. Charles Darwin posited a correlation between sea depth and isolation. He said, *"the deeper the sea, the more distantly related the fauna."* With eustatic sea level changes, those islands sharing a continental shelf would be frequently connected. Deep ocean trenches between islands would mean very rarely, if ever, would they be connected. Depth can be used as a marker of isolation. Based on this idea, and from the choices of: Java, Bali and Lombok, which two islands likely have a deep ocean trench between them?

_____ and _____

4. Wallace also noticed that one family of birds in particular, the Megapodes, was absent on Bali, Java, and islands to the north, yet present on Lombok and islands southward.

Megapodes like the Orange-footed guinea fowl (*Megapodius reinwardt*) are found on Lombok but not Bali and they have important characteristics that give clues to evolutionary histories.

The roughly two-dozen, mound building, chicken-like bird species are known for incubating their eggs on the ground. If the nest materials are comprised of rotting leaves, twigs and other debris, the male will either add material or take away material so as to allow for optimal incubation temperatures. Other sites such as volcanically warmed sands have been used as incubation sites as well.

Picture source: Wikipedia, "Megapodes"

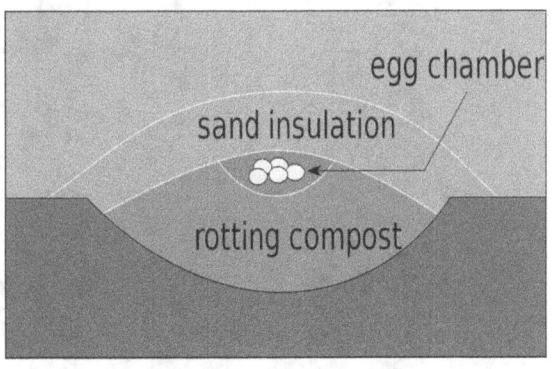

Based on the activity of ground dwelling, chicken-like birds, what creature(s) are likely to exist at very low densities, have small mobile populations, or even be completely absent on Lombok but present on Bali and Java?

_____

5. Since Megapodes exist on Lombok, the rest of the island chain south, New Guinea and Australia but they don't exist on Bali, Java or anywhere into Southeast Asia, what could you conclude about the dispersal of those creatures in your previous answer? Consider in relation to the movement between the Indo-Malayan and Australo-Malayan regions?

Wallace had determined that a definite boundary exists between Bali and Lombok, but then he had to determine if there was a continuation of the boundary elsewhere in the region. Is there a boundary between Borneo and Java? Perhaps between Borneo and Sulawesi? Or maybe between Sulawesi and New Guinea through the Moluccas?

Sulawesi had the defining evidence. Review the map below showing water depth (darkest is deepest) at the end of the last ice age 10200 years before present.

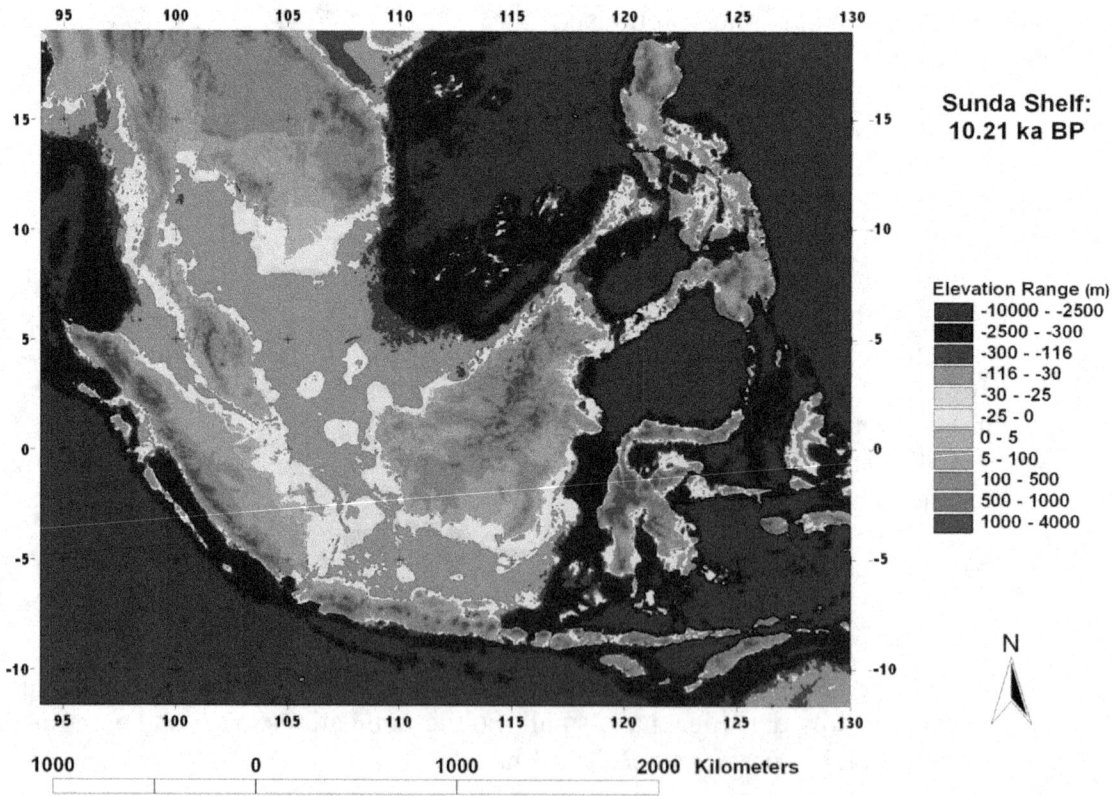

6. Would the islands of Sulawesi and Borneo have been connected? _____

7. Consider water depth around the entire island. Sulawesi's origin is likely?

Continental / Oceanic

8. Sulawesi is four times larger than Java and should therefore have greater species richness according to ....

_____

9. Sulawesi in actuality has half the number of mammals as that found on Java indicating that Sulawesi is likely this type of island.

Continental / Oceanic

10. While Sulawesi has fewer mammals than comparable islands, endemism is high, which suggests Sulawesi has been isolated for a:

Short period of time / Long period of time

11. Sulawesi has both placental and marsupial mammals. The marsupials would have arrived *before* / *after* the placental mammals, suggesting an arrival of more than _____ million years ago on Sulawesi.

12. The endemic marsupial civet known as the Sulawesi Palm Civet (*Macrogalidia musschenbroekii*) of Australasian origin indicates what about terrestrial animals in relation to this island:

    a) Terrestrial mammals rarely succeed in making it to oceanic islands
    b) The Sulawesi Civet was one of few mammals to successfully occupy the island
    c) The marsupial Civet likely arrived prior to mammalian evolution
    d) Sulawesi is not of Australasian origin
    e) All of the above

13. Topographic diversity, or *environmental heterogeneity*, is known to assist in speeding up evolutionary divergence.

    a. This process in which organisms diverge rapidly, particularly in environments that offer a range of new resources, is known as:

    _____

    b. Sulawesi has a low diversity of amphibians. In comparison to the Philippines however, it has a high diversity of Fanged Frogs. These have rapidly filled the ecological niches on Sulawesi that are normally occupied by other frog lineages on the Philippines. This suggests what about the topography of Sulawesi?

    _____

14. Sulawesi's unique composition of species initially had Wallace drawing a boundary between Sulawesi and Borneo. However, he vacillated indeterminately between there and having Sulawesi and the Moluccas astride an alternate boundary. In 1910 he reconsidered again, suggesting the line will forever be a topic of subjectivity.

Other lines were also proposed during the 19th Century. On the following page, find and briefly note the key evidence or reasoning behind the boundary lines drawn by others.

Salomon Müller (1846) _____

Thomas H. Huxley (1868) _____

The Sclater Line (1899) _____

Lydekker's Line (1896) _____

Further Recommended Reading:

van Oosterzee, P. (1997) *Where Worlds Collide: The Wallace Line*. Kew, Australia: Reed

APPENDIX A

Self-guided Field Trip
"extreme biodiversity gradients & refugia puzzle"

This field trip to the southwestern part of the Province of Alberta, Canada is meant to illustrate how predictable biogeographic patterns appear on the land, but also to witness unexpected anomalies on the current landscape. This is also tied into evidence of ancient patterns.

This exercise will guide you on investigating both current and ancient flora and fauna and to experience how biogeographic 'patterns and rules' shape what exists on the ground today.

Disturbance

Disturbances reset the balance of species composition and species richness.

Blitzkrieg Theory of Pleistocene Overkill

P.S. Martin's hypothesis that a geographically expanding front of prehistoric human colonists led to the extinction of mega fauna in North and South America.

Environmental Heterogeneity:

Greater topographic diversity is associated with greater species richness.

Isolation

Isolated populations often undergo rapid speciation including adaptive radiations.

Permafrost

Frozen ground and subsoil that has not thawed for a minimum of two years.

Refugia

A geographic place that has remained unaltered by a climatic change, often in the form of a glacial episode, affecting surrounding regions and that therefore forms a haven for fauna and flora. Post-glacial refugia survivors often exist as relic populations.

Chinook

An adiabatically warmed air mass, leeward side of mountain ranges, often resulting in rapid temperature changes and strong winds. Can result in an inconsistent snow cover through the winter season on some landscapes.

Site#1

ST. MARY PROVINCIAL RECREATION AREA, ALBERTA

The 'Blitzkrieg Theory of Pleistocene Overkill' has gone through many ups and downs as a potential hypothesis to explain the mass extinction of North American mega fauna at the end of the Pleistocene and beginning of the Holocene Epochs. The initial proposal suggested early humans had played a significant role in mega fauna extinctions through overhunting. However, archaeological kill-sites were showing an absence of horse (Equus) remains, thus suggesting horse was strangely not on the hunters' list of prey. This led to early conclusions that humans did not have the ability to play a large role in affecting mega fauna populations.

However, in 1996 local schoolteacher Shayne Toman, his friend and wife, discovered significant remains of horse and camel in kill-sites at the St. Mary Provincial Recreation Area, in southern Alberta. As of 2015, it remains the only direct evidence in North America of early humans hunting camel and horse around 13,300 years ago. This find in 1996 re-established the Blitzkrieg Theory of Pleistocene Overkill as a potential explanation for mega fauna extinctions. However, for other reasons, today the Blitzkrieg Theory still remains contentious in the academic field.

The ideal time to visit this site is in early fall during a dry year, when water levels are low and flora are still easy to identify.

Wally's Day Use area is the ideal place to do this exercise. It can be accessed off Highway 505 near the dam. The short access road is gravel and easily passable by all vehicles.

Take note how the day use area is mostly 'natural' but is surrounded by the monoculture of farmland, flat topography, and bordered by an artificial reservoir where a natural riparian zone used to exist. Today, this 'island zone' stands in contrast to its prior landscape. At the end of the Pleistocene this regional landscape would have been a newly ice-free terrain of grasslands occupied by bison, camels, horses, muskox, caribou, mammoths and some predators such as Grizzly bears and sabre-tooth cats.

Explore the Day-Use area and answer the following questions. If you are lucky enough to come at the right time of year, your wanderings to the waters edge may reward you with finds of bones and even arrow points. Please enjoy the artifacts but leave them in place.

Sources

Grayson, D.K. (1991) Late Pleistocene Mammalian Extinctions in North America: Taxonomy, Chronology and Explanations. Journal of World Prehistory 5: 193-231.

Kooyman, B., Newman, M.E., Cluney, C., Lobb, M., Tolman, S., McNeil, P., and Hills, L.V. (2001), Identification of Horse Exploitation by Clovis Hunters Based on Protein Analysis. *American Antiquity*, 66: 686-691

1. Identify at least 12 plants. Try to find more *if possible*.

COMMON/SCIENTIFIC NAME	NATIVE	
_____	NO	YES
_____	NO	YES
_____	NO	YES
_____	NO	YES
_____	NO	YES
_____	NO	YES
_____	NO	YES
_____	NO	YES
_____	NO	YES
_____	NO	YES
_____	NO	YES
_____	NO	YES
_____	NO	YES
_____	NO	YES
_____	NO	YES
_____	NO	YES
_____	NO	YES

2. Do you see any large mammals around? Consider how late Pleistocene/early Holocene horse *(Equus conversidens)*, bison *(Bison bison antiques)*, helmeted musk oxen *(Bootherium bombifrons)* and caribou *(Rangifer tarandus)* would have shaped the vegetation at around 13,000 years ago. What would the vegetation likely consist of considering that horses, caribou, and musk oxen are grazers more than browsers? How would this be different than today?

3. What might you conclude about the climate at that time?

Site#2

SOUTHWEST ALBERTA & THE CROWSNEST PASS AREA

After visiting St. Mary Reservoir, drive west on Highway 505. Once you reach Highway 6, drive north and look for a place to pull over on the side of the road. Many places exist for you to count the natural floral diversity within a reasonable area.

3. After compiling your list of plants answer why this area (Waterton National Park to the Crowsnest Pass) contains not only higher diversity than you found at St. Mary Reservoir but also contains the highest biodiversity in the entire Province of Alberta?

4. How do the factors that explain higher biodiversity in this area compare to the much lower diversity at St. Mary Reservoir? What are the differences?

Notes:

Site#3

PLATEAU MOUNTAIN ECOLOGICAL RESERVE

The previous two stops illustrate how biogeographic patterns are often quite predictable based on underlying influences such as topography, isolation and time. This next stop will introduce you to an anomaly—a refugia puzzle.

In 1991, the Alberta Provincial Government officially designated Plateau Mountain as an ecological reserve set aside for ecological preservation, scientific research and limited recreation activities. With a 4% grade that defines the top of the mountain, it is quite rich in floral diversity overall with approximately five hundred vascular plant species, seven of which are rare. Alpine and Tundra flora, while both are found in cold environments, are often quite distinct from one another and are not usually found together. Plateau Mountain however, contains both Alpine flora and an unusually abundant collection of Tundra flora. While both are associated with cold condition, Tundra species are specifically associated with permafrost. Plateau Mountain, unlike surrounding mountains, has a very deep layer (100ft) of permafrost on its flat top.

This exercise will have you answer why tundra species are found on this particular mountain but not elsewhere in the region. Why is this place such a species anomaly?

While you can access the mountain itself with a day hike and enjoy a wonderful view of the surrounding area, it is also possible to simply view Plateau Mountain looking west from Highway 22 near Chain Lakes Provincial Park

Sources

Alberta Environment (2000), Plateau Mountain Ecological Reserve Management Plan. Canmore, Alberta: Natural Resources Service.

"Plateau Mountain: A Case Study of Alpine Permafrost in the Canadian Rocky Mountain", (with R.J.E. Brown). *Proc Third Internat. Conf. On Permafrost*, Edmonton, Alberta, vol.1, 1978, 385-391.

"Permafrost Distribution Along the Rocky Mountains in Alberta" (with the late R.J.E. Brown).1982. The Roger J. E. Brown Memorial Volume, *Proceedings of the 4th Canadian Permafrost Conference, Calgary, Alberta*, H.M. French, Ed. National Research Council, Ottawa, pp.59-67.

"Distribution and Zonation of Permafrost Along the Eastern Ranges of the Cordillera of North America", *Biuletyn Peryglacjalny*, vol. 31, 1986, 107-118.

"Continentality Index: Its Uses and Limitations When Applied to Permafrost in the Canadian Cordillera", *Journal of Physical Geography*, vol. 10. 1989, 268-282.

"The periglacial environment of Plateau Mountain: An overview of current periglacial research." *Polar Geography*, vol.

21, 1997, 113-136 [with A. Prick].

"A relict of Late Quaternary permafrost on a former nunatak at Plateau Mountain, Alberta." [in Russian]. *Earth Cryosphere*, vol. 1. No. 4, 1997, 20-27.

"Relict Late Quaternary permafrost on a former nunatak at Plateau Mountain, S.W. Alberta, Canada." [English & Polish]. *Biuletyn Peryglacjalny*, vol. 36, 1998, 47-72.

"Long-Term Air and Ground Temperature Records From the Canadian Cordillera", *Proc. 5th Canadian Permafrost Conference*. National Research Council of Canada. Universite Laval, Collection Nordicana #54, 1990, pp. 151-157.

Background Research & Data Review

Examine the following background information on the research and conditions connected to this unique mountain and its unusual collection of flora before answering the last question.

Source: Circumpolar Active Layer Monitoring component of Global Terrestrial Network for Permafrost

Year Started: 1976

Responsible individual(s): Stuart. A. Harris

Latitude and longitude: 50°13'N 116°31'W

Slope and aspect of site(s): 4% slope N

Landform or geomorphological description of area occupied by site(s): N. end of flat mountaintop with 4% slope N.

Predominant texture of soil (sand, gravel, peat, etc.): Shattered rock and loess

Vegetation classification: Alpine tundra 20% cover

Method to determine active layer thickness: Interpolation of ground temp. profile

Information on permanent temperature and/or moisture installations (probes, recorders, depth of sensors, depth of permafrost boreholes, etc.): 14 YSI 44033 thermistors to 100'depth. These have been logged by data logger twice daily since 1993. Prior to this, it was logged manually

5. The 'Flame-coloured Lousewort' (*Pedicularis flammea*) is a common tundra species with a habitat range normally found across the Canadian Arctic, Iceland and Greenland, but is also found on Plateau Mountain as a rare species. Why might this isolated population be found here, far south of its present tundra range? To answer, use the following points to guide you through to an answer.

Points to consider:

o   When would tundra plants have been commonly found across this part of North America, during what geologic time interval?

o   What conditions would need to remain favorable, for tundra plants to remain atop Plateau Mountain when global climatic conditions changed? What ground condition is associated with tundra plants?

o   Consider what helps or hinders the ideal 'ground condition' for tundra plants?

o   What would allow for a favourable 'ground condition' to exist atop this specific location (Plateau Mountain) but not in surrounding locations?

o   What is unique about Plateau Mountain? Consider two components: the shape of it, and the specific location in relation to regional weather patterns.